I0817395

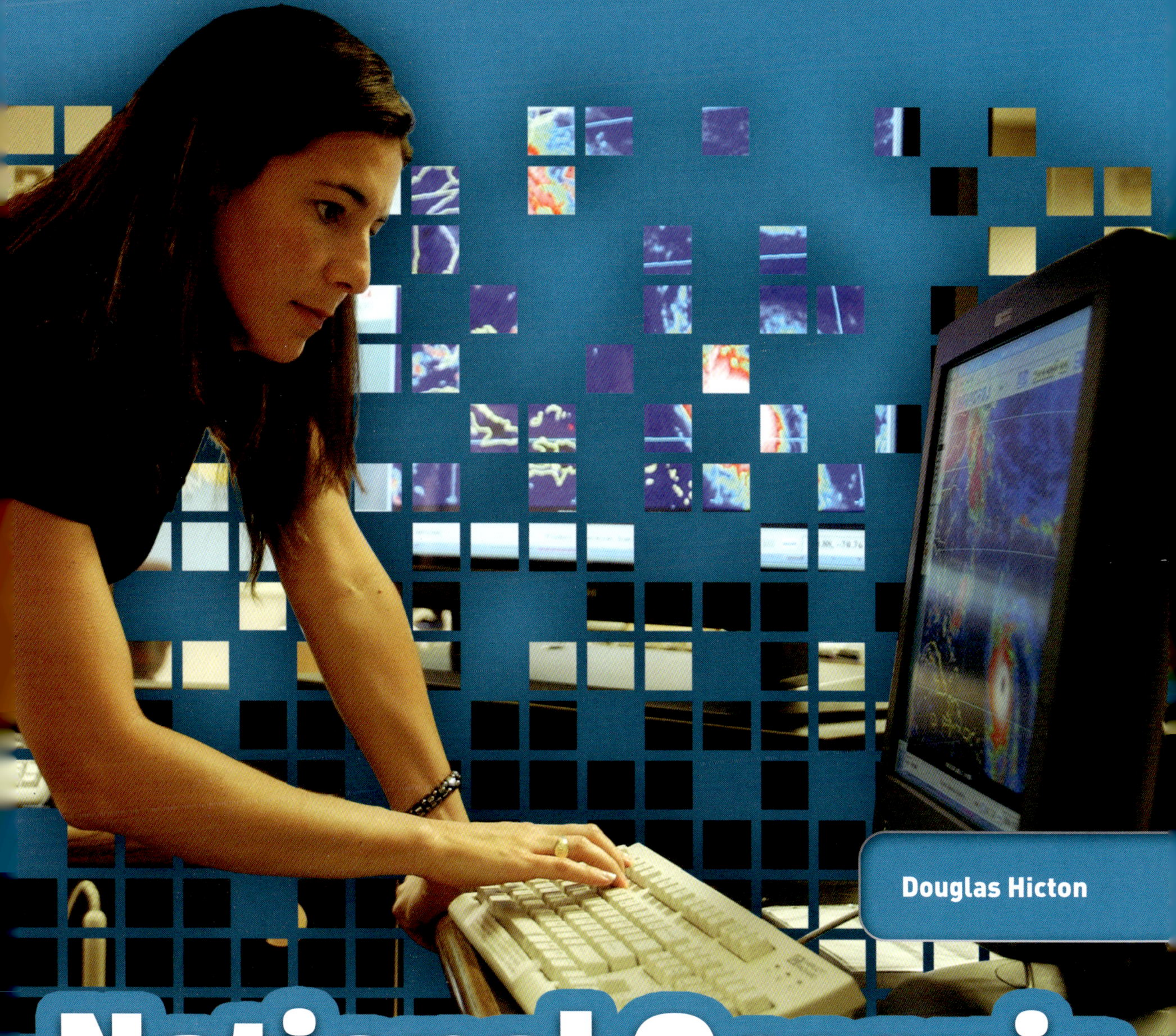

Douglas Hicton

National Oceanic and Atmospheric Administration

POWER • AUTHORITY • GOVERNANCE

LIGHTBOX
openlightbox.com

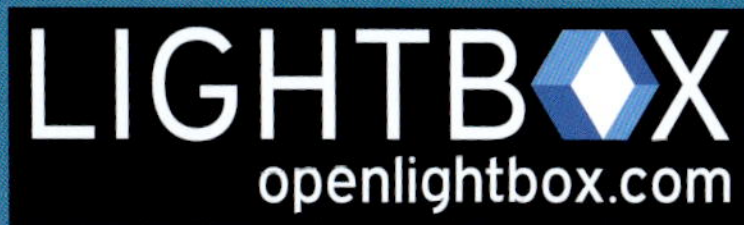

Go to
www.openlightbox.com
and enter this book's
unique code.

ACCESS CODE

LBXT5385

Lightbox is an all-inclusive digital solution for the teaching and learning of curriculum topics in an original, groundbreaking way. Lightbox is based on National Curriculum Standards.

LIGHTBOX SUPPLEMENTARY RESOURCES

SHARE
Share titles within your Learning Management System (LMS) or Library Circulation System

CURRICULUM
Find national and state curriculum correlations

CITATION
Create bibliographical references following APA, CMOS, and MLA styles

STANDARD FEATURES OF LIGHTBOX

AUDIO High-quality narration using text-to-speech system

ACTIVITIES Printable PDFs that can be emailed and graded

SLIDESHOWS
Pictorial overviews of key concepts

VIDEOS Embedded high-definition video clips

WEBLINKS Curated links to external, child-safe resources

TRANSPARENCIES
Step-by-step layering of maps, diagrams, charts, and timelines

INTERACTIVE MAPS
Interactive maps and aerial satellite imagery

QUIZZES Ten multiple-choice questions that are automatically graded and emailed for teacher assessment

KEY WORDS
Matching key concepts to their definitions

This title is part of our Lightbox digital subscription

Lightbox Grades 3–5 Subscription
ISBN 978-1-5105-5424-5

Access hundreds of Lightbox titles with our digital subscription. Sign up for a **FREE** subscription trial at **www.openlightbox.com/trial**

POWER • AUTHORITY • GOVERNANCE

National Oceanic and Atmospheric Administration

CONTENTS

Introduction

Citizens of the United States determine the nature of their government and its various parts. That is the definition of "popular sovereignty." The people elect representatives who then form governments. In turn, governments create agencies to help the people. Some agencies have proven their worth many times.

"From the surface of the Sun to the depths of the ocean floor," the National Oceanic and Atmospheric Administration (NOAA) collects data on weather and climate. It then shares this data in the form of weather forecasts. Forecasts are useful for people in the path of weather disasters. These disasters cause billions of dollars in damage every year and threaten many lives, making this element of NOAA extremely important.

NOAA's headquarters are located in Silver Spring, Maryland, just north of Washington, DC.

NOAA also has an important role related to the **environment**. The agency helps to conserve and manage **ecosystems** on U.S. coasts and in the ocean beyond. By doing so, NOAA can ensure that these areas are protected from pollution and other threats.

The ocean is a source of food through fishing. It is also an international highway for shipping. This means that the ocean is not just a vital part of the environment, but of the U.S. economy as well. For this reason, NOAA does its work within the Department of Commerce.

NOAA collects **20 terabytes** of data every day. This is **twice the volume** of data of all the printed works in the Library of Congress.

NOAA's polar geostationary satellites orbit about **22,240 miles** (35,800 kilometers) above the planet.

The U.S. Department of Commerce was created on **February 14, 1903**. In addition to NOAA, it oversees agencies including the **Census Bureau** and the **Patent and Trademark Office**.

Origins of NOAA

The Survey of the Coast was the oldest scientific agency of the American government, dating from 1807. The Weather Bureau was founded in 1870, while the U.S. Commission of Fish and Fisheries became the first federal conservation agency in 1871. The first two agencies were combined into the Environmental Science Services Administration (ESSA) in 1965. In 1969, a commission of leading scientists recommended that the ESSA, Fisheries, and several other agencies be combined into a new organization, NOAA.

Both the Department of Commerce and the Department of the Interior wanted to oversee NOAA. At the time, the ESSA was already in the Department of Commerce. However, conservation of natural resources was the main responsibility of the Department of the Interior. The Department of the Interior was also the home of the U.S. Geological Survey, so it seemed like the more sensible place for the new agency. By the spring of 1970, it looked as if the White House would choose that department. However, after President Richard Nixon found himself in conflict with his secretary of the interior, Wally Hickel, this decision was changed. As a result, Nixon signed NOAA into being on October 3, 1970, as a bureau of the Department of Commerce.

NOAA launched its first weather satellite, NOAA-1, into orbit the same year the agency was founded.

Branches of Government

NOAA reports to the U.S. secretary of commerce. The secretary is a member of the federal cabinet, which reports to the head of the **executive branch** of government, the president. Before taking office, both the commerce secretary and the NOAA administrator must be confirmed by the Senate, which is the upper house in the **legislative branch**, or Congress.

Meanwhile, the lower house of Congress, the House of Representatives, has what is often called "the power of the purse." This means it alone controls how the national government raises taxes and spends money. Each year, the president presents budget proposals to Congress. Both houses must pass the budget.

Finally, federal courts, in the **judicial branch**, interpret laws and **regulations**. If laws do not follow the **Constitution**, the courts can strike them down. The framers of the Constitution put such "checks and balances" in place to ensure that no single branch of government becomes too powerful.

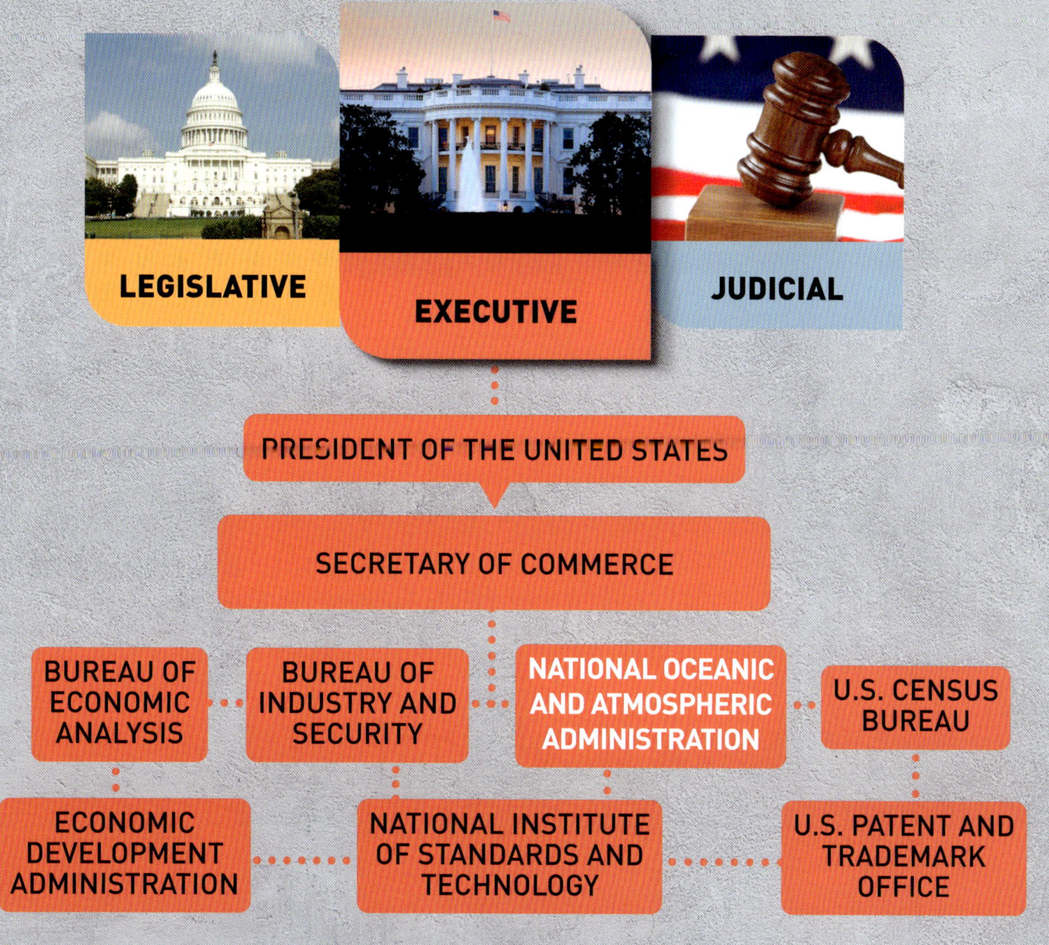

Purpose of NOAA

The mission of NOAA is, in part, "to understand and predict changes in climate, weather, oceans, and coasts." The organization shares this information to help keep people safe. It also uses it to protect the nation's marine ecosystems. The agency is divided into six offices.

NOAA's offices employ 12,000 personnel worldwide, including almost 6,800 scientists and engineers.

Weather disasters cost the United States more than 350 lives and $140 billion every year. However, this toll might be much higher without the National Weather Service (NWS), a NOAA office. When severe weather threatens lives and property, warnings from the NWS give people time to evacuate or prepare.

U.S. ports handle goods worth $1.5 trillion every year. All ships rely on navigation charts and water level information from the National Ocean Service (NOS). This information keeps the ships safe. Many private businesses also make use of reliable NOS data.

Commercial fisheries provide jobs for 1.2 million Americans and food for millions of people around the world. The National Marine Fisheries Service (NMFS) maintains marine habitats and safe and **sustainable** sources of seafood. NMFS works with regional fisheries to set catch limits, reduce **bycatch**, and ensure regulations are followed.

NOAA research scientists at Oceanic and Atmospheric Research (OAR) assemble and analyze data. They use the data to better understand problems such as climate change, natural disasters, and ocean acidification.

The National Environmental Satellite, Data, and Information Service (NESDIS) operates NOAA's nine weather satellites. These satellites have tracked hurricanes, mapped flooded areas, and helped create scientific models to predict the spread of wildfire smoke. Finally, the Office of Program Planning and Integration (PPI) coordinates the work of the other five offices, both among themselves and with other agencies.

Increasing Disasters

In 2020, wildfires raged across much of the western United States. The fire season in California was particularly intense. Wind, high temperatures, and low humidity created perfect conditions for fires. All it took to start one was a spark, often caused by fireworks or a campfire. Other sources of sparks included the unusually large number of thunderstorms that year.

NOAA reports that the United States had an average of six $1-billion weather or climate disasters every year from 1980 to 2013. Since 2014, the yearly average has become about 12 events. The more than 270 disasters since 1980 have caused damage in excess of $1.8 trillion. This increase in annual disasters means that NOAA and the role it plays in managing them has become even more important over time.

NOAA Through the Years

NOAA was pieced together more than 50 years ago, but the history of its individual components goes back much farther. Over two centuries, NOAA's responsibilities have evolved, as have its methods and equipment.

February 10, 1807

President Thomas Jefferson authorizes the Survey of the Coast. This agency is meant to provide accurate nautical charts to ensure the safety of maritime commerce. It is later renamed Coast Survey in 1836.

February 9, 1870

President Ulysses S. Grant creates the national Weather Bureau, to be run by the U.S. army. The bureau is moved to civilian government by 1891.

February 9, 1871

Congress creates the U.S. Commission of Fish and Fisheries (CFF) as an **independent agency**.

1878

Coast Survey is renamed U.S. Coast and Geodetic Survey (C&GS). Over time, its mission expands to include most of the physical sciences, including oceanography, **seismology**, and tide prediction.

September 8, 1900

The Great Galveston storm strikes the United States. It is one of the deadliest natural disasters in U.S. history.

1903

CFF becomes the U.S. Bureau of Fisheries, in the Department of Commerce and Labor. In 1939, it moves to the Department of the Interior.

1948

C&GS establishes the Seismic Sea Wave Warning System in Honolulu, Hawaii. It is renamed the Pacific **Tsunami** Warning System in 1965.

April 1, 1960

TIROS-1, the world's first weather satellite, is launched. Over the next two and a half months, it takes 19,389 usable pictures.

October 3, 1970

President Nixon issues an **executive order** establishing NOAA.

1979

After 25 years of naming hurricanes only after women, the United States announces it will begin alternating between male and female names.

1980

NOAA establishes the National Undersea Research Program. Today, it has research centers in places including Hawaii, Alaska, and Florida.

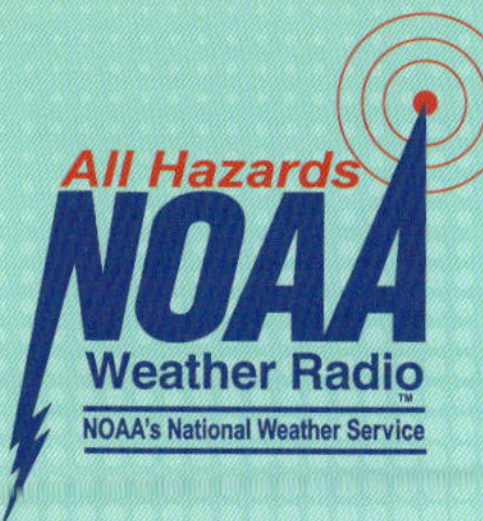

2003

NOAA Weather Radio expands to more than 800 transmitters nationwide. It can now reach 95 percent of the American public.

2021

NOAA announces that 2020 was Earth's second-hottest year on record, after 2016.

2022

NOAA celebrates the 50th anniversary of Congress passing legislation meant to protect U.S. oceans and coasts.

NOAA Issues

NOAA is a huge federal agency, and managing both the atmosphere and the oceans are important. However, NOAA is just one of many agencies in Commerce, and Commerce is just one of 15 Cabinet departments, each of which requires money in order to carry out its responsibilities.

Many of the tools used by NOAA, such as weather satellites, are expensive. This can cause problems when determining budgets, as there is only so much money to go around. In 2010, NOAA received $4.7 billion from Congress, $1.3 billion of which was for acquisition, or buying equipment and supplies. Of the acquisition budget, 90 percent went toward upgrades to NOAA's aging satellite systems.

Two years later, NOAA's acquisition budget had risen to $1.8 billion, $1.7 billion of which went to more satellites. This took much-needed funding away from fisheries management, ocean monitoring, and pollution response. Each year from 2017 to 2020 saw further cuts to NOAA's budget. Due to these changes, the 2018 budget proposed the shutdown of most climate research programs.

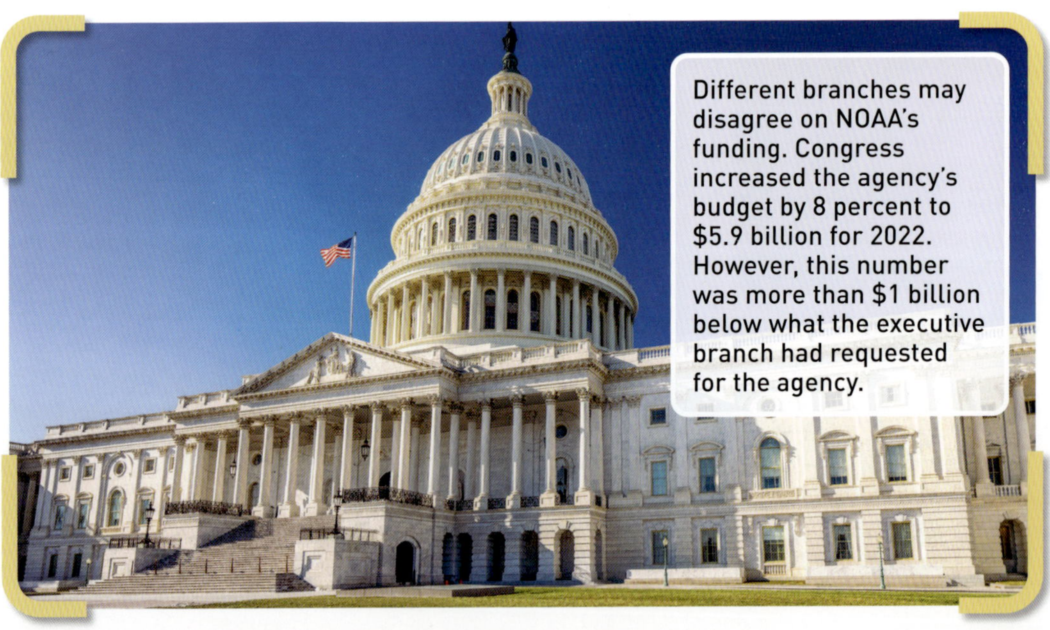

Different branches may disagree on NOAA's funding. Congress increased the agency's budget by 8 percent to $5.9 billion for 2022. However, this number was more than $1 billion below what the executive branch had requested for the agency.

NOAA has also had political controversies. In 2017, the CEO of a company that used data from NWS satellites was a nominee to head NOAA. Some senators were concerned about possible conflicts of interest. By November 2019, he had withdrawn his nomination.

Hurricane Dorian

As Hurricane Dorian sped westward across the Atlantic in 2019, NOAA scientists were monitoring it closely and providing updates. They expected the powerful storm to pummel the southeastern United States. The president issued a statement warning several states, including Alabama, about the hurricane. However, the Birmingham office of NWS quickly issued a statement of its own saying that Alabama would not be affected.

Eventually, NOAA released a new statement, saying that the original NWS statement had been worded with too much certainty and that Alabama could be affected. However, some felt that this new statement was not scientifically motivated. Ultimately, an independent panel ruled that senior NOAA officials had violated the agency's code of scientific integrity with the new statement.

Hurricane Dorian caused more than $1.6 billion in damages in the United States alone.

Key Figures in NOAA

An uncountable number of men and women have lent their talents to NOAA over two centuries. Many have left a lasting mark on the organization through their leadership.

Alexander Dallas Bache

Alexander Dallas Bache (1806–1867) was the second superintendent of the Coast Survey. He turned it into one of the world's top scientific organizations. A great-grandson of Benjamin Franklin, Bache thought science should serve the nation's commercial and defense interests.

Evelyn Fields

Evelyn Fields (1949–) was the first African American woman to join NOAA Corps, the agency's uniformed division. In her 30-year career, she was the first female commander of a federal ship, the first female NOAA rear admiral, and the first female and first African American director of NOAA Corps.

Kathryn Sullivan

Kathryn Sullivan (1951–) was the first American woman to walk in space. After three shuttle missions, she left NASA to become NOAA's chief scientist. Sullivan later served as deputy administrator of NOAA before taking on the role of administrator until early 2017.

The Great Galveston Storm

HISTORICAL CASE STUDY

Isaac Cline was chief **meteorologist** at the U.S. Weather Bureau (USWB) station in Galveston. The city was located at the northeast end of Galveston Island, on Texas's Gulf coast. On Friday, September 7, 1900, the USWB in Washington sent Cline a telegram. A tropical storm was moving over Cuba toward Florida.

Cuban meteorologists had observed the storm since Monday. It was growing into a large hurricane. However, the USWB did not believe the forecast. There was tension between the United States and Cuba after the Spanish-American War. USWB director Willis Moore would not let the Cubans send weather warnings to American stations. All reports had to go through him. He sent no hurricane warning to Galveston on September 8.

Driving rain and 20-foot (6-meter) tides flooded Galveston Island, uprooting houses, which disintegrated in the churning water. However, according to some accounts, Cline had decided to warn people based on his own observations. Many people have credited him with saving thousands of lives.

The Great Galveston Storm killed between 6,000 and 12,000 people.

Careers in NOAA

Physical Oceanographer

Physical oceanographers study waves, currents, tides, and coastal erosion. Additionally, they examine how the ocean and the atmosphere interact to influence weather and climate. NOAA also has biological, chemical, and geological oceanographers. All oceanographers must have a keen understanding of biology, chemistry, geology, and physics.

Meteorologist

Meteorology is the scientific study of the atmosphere, especially as it relates to weather forecasting. NOAA meteorologists need a working knowledge of climatology, advanced **thermodynamics**, calculus, chemistry, **aeronomy**, computer science, electricity and magnetism, light and optics, and meteorology.

Civilian Mariner

NOAA has a fleet of science and survey ships commanded and managed by uniformed officers from NOAA Corps. It also relies on civilian officers and mariners. Civilian mariners are the fleet's backbone. Any single mariner might do several jobs, from collecting scientific data to servicing buoys, feeding the crew, and maintaining the ship.

Hurricane Hunter

NOAA Corps employs trained pilots to fly aircraft into hurricanes. The job can be both exciting and terrifying. An onboard meteorologist collects data and directs the pilot where to go. Hurricane hunters seek the zero-wind center of the storm's eye, the area of lowest atmospheric pressure.

NWS hurricane hunters have **lost only one plane** in their history. *Snowcloud Five* flew into Hurricane Janet on **September 26, 1955**, and was never seen again.

The **NOAA Corps fleet** includes **15 ships** and **9 airplanes.**

STEM stands for **Science**, **Technology**, **Engineering**, and **Math**. Since 2005, **NOAA's Environmental Literacy Program** has **awarded 145 grants** toward environmental and STEM education, totaling $81 million.

Tools of the Trade

NOAA needs many different tools to do its work. Some of these tools are used for specific jobs. Other tools are more common. They are used by most NOAA workers.

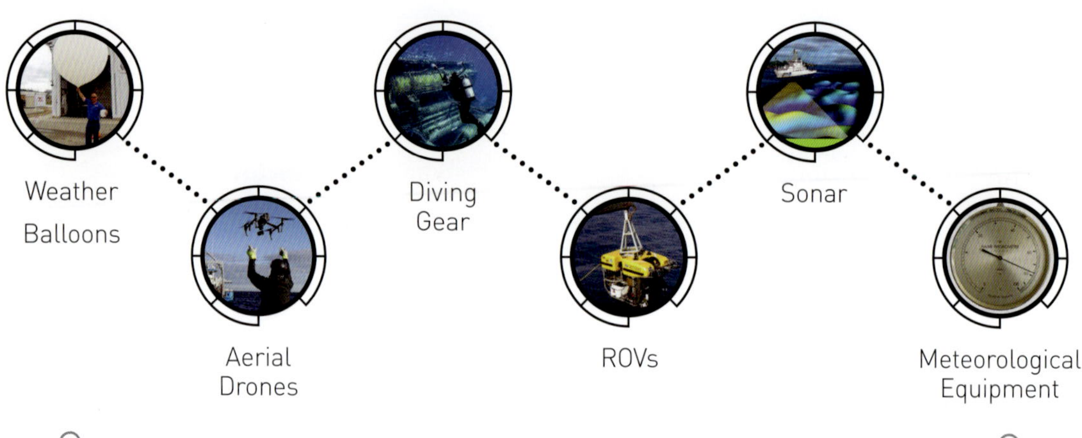

Weather Balloons

Every 12 hours, 92 NWS offices release helium- or hydrogen-filled weather balloons. Each carries a sensitive piece of equipment called a radiosonde. It measures temperature, humidity, and atmospheric pressure. The balloons can rise as high as 100,000 feet (30,500 m) before they burst. NWS has been regularly using weather balloons since the 1930s.

Aerial Drones

NOAA's remote-controlled aircraft, or drones, can be used to survey the extent of flood damage from high in the sky. Some can get close enough to marine mammals to film them or even sample their breath. Drones can chart sea habitats, monitor hurricanes, and assess fishery stocks. They also assist with emergency responses.

Diving Gear

All NOAA divers are trained to wear standard gear in a consistent way. Familiarity with their life support equipment makes divers safer. It also lets them concentrate on their tasks. Divers take care of their own gear, which includes scuba tanks, reserve air supply systems, diving regulators, depth and pressure gauges, wetsuits, drysuits, masks, and fins.

ROVs

Remotely Operated Vehicles (ROVs) explore ocean areas too deep for humans to dive safely. NOAA scientists operate them remotely from ships. Most ROVs are equipped with cameras and lights to transmit images and video to the surface. Other equipment, such as robotic arms, may be attached to them. ROVs can be as small as a laptop or as large as a truck.

Sonar

Active sonar allows people to detect objects underwater. This tool emits sound pulses that bounce off objects and echo back. The strength of the echo depends on how far the object is from the pulse's source. NOAA uses active sonar to map the contours of the seabed. It also uses passive sonar. This can detect sounds made by whales and other sea creatures.

Meteorological Equipment

The NWS uses many kinds of traditional equipment. Thermometers measure temperature. Anemometers measure wind speed. Barometers measure atmospheric pressure. Hygrometers measure humidity. Rain gauges measure rain. These instruments can be found in any NWS weather station.

NOAA in the United States

NOAA influences the lives of Americans every day. Whenever someone checks the weather, sees a weather record, or receives a warning about a weather disaster, the information has come from NOAA.

1

Boulder, Colorado

Strong winds called Chinooks tore through the Boulder area in January of 1982. On the 17th, NOAA's Environmental Research Laboratory reported winds of more than 100 miles (160 km) per hour. The Chinook damaged buildings, snapped utility poles, and destroyed planes at the airport.

2

Death Valley, California

Death Valley is long, narrow, and surrounded by very steep mountain walls that trap the Mojave Desert's heat. On August 16, 2020, weather station thermometers read 129.9 degrees Fahrenheit (54.4 degrees Celsius) at the area's Furnace Creek Visitor Center. NOAA noted that this may be the hottest air temperature ever reliably recorded.

CANADA
UNITED STATES
MEXICO
Washington
Montana
Oregon
Idaho
Wyoming
Nevada
Utah
Colorado
California
Arizona
New Mexico
Pacific Ocean

LEGEND

- Land (USA)
- Land (Other)
- Water

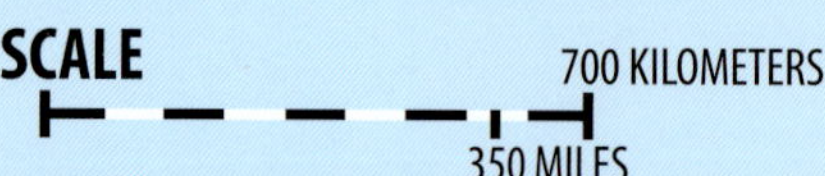

Fargo, North Dakota

A massive tornado plowed through Fargo on June 20, 1957, with the energy of 25 atom bombs. The tornado was spawned in a **supercell** storm along with four others. It caused massive destruction, but only a dozen lives were lost. The weather office had been able to warn people to flee in time.

Cape Hatteras, North Carolina

The USS *Monitor* fought for the Union during the Civil War. It sank in a winter storm off Cape Hatteras in late 1862. The shipwreck became an artificial reef, home to a large amount of sea life. However, nobody knew where it was until 1974. On January 30, 1975, the area around the wreck was protected as the Monitor National Marine Sanctuary. It was the first of NOAA's 14 national marine sanctuaries.

NOAA in the World

The first part of NOAA's mission is understanding and predicting changes in climate, weather, oceans, and coasts. As events in one area of the ocean can affect another area halfway around the world, this mission extends beyond the political boundaries of the United States. The second part of NOAA's mission is sharing knowledge and information. NOAA shares data freely and gets back three times as much in return.

In late 2020, NOAA entered into a new agreement with the Japan Aerospace Exploration Agency (JAXA). The goal of this agreement is to improve global weather forecasts, something that is important to both nations. The United States broke a record for the number of billion-dollar weather disasters in 2020, and storms called typhoons have always been a problem facing Japan.

The United States joined the International Meteorological Organization as far back as 1873. This group was replaced by the World Meteorological Organization in 1950.

NOAA works with groups around the world to protect marine life. The International Coral Grant Program, administered by NOAA, supports the development of national marine protected areas. Through international fisheries organizations, NOAA helps to promote science-based management of fish stocks. Its Office of Law Enforcement works with international partners to combat illegal, unreported, and unregulated fishing.

The NOAA Coral Reef Conservation Program awards grants to support coral reef conservation efforts. These include domestic U.S. grants and international grants for areas around the world.

The **United States imports** more than **60 percent** of its **seafood** from **overseas.**

The **World Meteorological Organization** has **193 member states** and **territories**. It is an agency of the United Nations.

The **International Coral Reef Initiative declared 2018** as the **third International Year of the Reef**. The first two were **1997** and **2008**.

NOAA Today

Almost three-quarters of Earth is covered by the ocean. It provides half of the planet's oxygen and has sustained civilizations for thousands of years. However, vast areas of the ocean remain uncharted. Only 47 percent of U.S. territorial waters have been properly mapped by today's standards. Less than 10 percent of the ocean has been explored.

NOAA has been taking steps to fill in the gaps. In partnership with various agencies and companies, NOAA has set up the Argo Global Ocean Array. About 4,000 floats distributed around the world—including in U.S. waters—measure ocean temperature, **salinity**, and other variables on the surface and far below. Approximately 140,000 measurements are taken every year. NOAA scientists hope these will tell them how changes in the bottom half of the ocean influence weather, climate, and rises in sea level.

NOAA is the principal U.S. agency devoted to the study of climate change. In February 2021, NOAA warned that the previous month was the seventh-warmest January on record. The agency also reported that the warmth had led to below-average levels of sea ice around the north and south poles. NOAA noted that the month was also the 433rd consecutive month with temperatures above the 20th-century average.

Each Argo float costs between $20,000 and $150,000. The floats are designed to sink in the water and then return to the surface over 10-day periods.

Japan Earthquake and Tsunami

MODERN CASE STUDY

On March 11, 2011, a magnitude 9.0 earthquake struck the Pacific Ocean about 80 miles (130 km) off the northeastern coast of Honshu, Japan. The 33-foot (10-m) tsunami that followed 20 minutes later flooded communities as far as 6 miles (10 km) inland. The tsunami also spread eastward to the Americas and south to Antarctica. NOAA maintains a large network of buoys in the Pacific. The buoys are connected to pressure sensors on the seabed. Just 25 minutes after the quake, the first buoy station registered the tsunami and relayed information to NOAA's Pacific Tsunami Warning Center in Hawaii. This was early enough notice for the people there to seek safety.

The tsunami severely damaged Japan's Fukushima Daiichi nuclear power plant. For three weeks, radioactive water leaked into tunnels that led to the ocean. A year after Fukushima, NOAA Fisheries scientists examined 50 tuna caught off the west coast of North America. The tuna tested positive with enough radioactivity for people to know the fish had traveled from Japan. Unexpectedly, the disaster helped NOAA understand the dynamics of tuna migration.

Shortly after the disaster, tens of thousands of people living near the Fukushima Daiichi nuclear power plant had to be evacuated due to radiation in the air.

The Honshu tsunami caused more than 450,000 people to lose their homes.

NOAA Looking to the Future

NOAA continues to embrace innovation. The agency intends to develop new robots and **autonomous** underwater vehicles that can collect ultra-high-resolution information. New technology will also advance NOAA's drone systems and artificial intelligence. These tools will be necessary for mapping U.S. waters. For example, as of 2021, only about 5 percent of the waters in the Great Lakes area have been mapped. In 2020, NOAA announced that its scientists would also join cruises on the Great Lakes to conduct research on changes in weather, climate, and ecosystems. Guests on the ships will be able to engage with the scientists in the on-board lab or participate directly in citizen science programs.

NOAA has announced it will completely map U.S. deep water by 2030 and near-shore waters by 2040 using data from a number of different sources, including the agency's own ships.

In 2020, NOAA released a new plan outlining its priorities for research and development until 2026. The priorities included reducing the impact of severe weather on society, along with sustainable use of ocean resources. NOAA wrote this plan based partly on suggestions from the general public. After the first draft was written, NOAA once again invited people to submit their comments before it wrote the final draft. In the future, similar steps to take public opinion into consideration could further enhance the role of popular sovereignty in NOAA's operations.

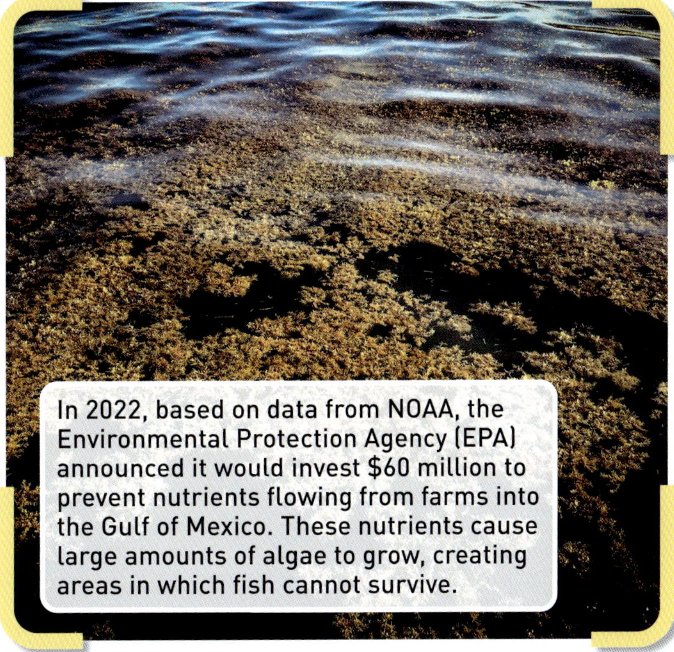

In 2022, based on data from NOAA, the Environmental Protection Agency (EPA) announced it would invest $60 million to prevent nutrients flowing from farms into the Gulf of Mexico. These nutrients cause large amounts of algae to grow, creating areas in which fish cannot survive.

For 2023, NOAA's proposed budget includes expanding the organization's ability to monitor Earth's changing climate and develop tools and strategies to help respond to climate-related disasters.

ACTIVITY ★★

Create a Policy Paper

NOAA has been in the Department of Commerce for all of its history. Some people see this as a basic conflict between business and environmental concerns. In 2011, federal employees were asked to submit ideas on how to make government more efficient. Some felt moving NOAA to the Department of the Interior would help with this issue. In 2012, the idea of transferring NOAA to Interior became part of a proposal to consolidate several agencies and save $3 billion. However, many members of Congress were against the idea, as were some environmental groups. The proposal ultimately went nowhere.

Develop your own thoughts about which cabinet department, if any, is the best fit for NOAA. Write a policy paper that summarizes your opinions.

Step 1:

Answer the following questions to help you develop your opinions.

1. Should NOAA be in the Department of Commerce, the Department of the Interior, or in a different department entirely? Give reasons for your answer.
2. Is there a conflict between NOAA's focus on the environment and Commerce's emphasis on business and trade? Why or why not?
3. If Interior took charge of NOAA, how would this change its role compared to other cabinet departments?
4. How would transferring NOAA to either Interior or another department affect businesses in the country? Would they support or oppose such a transfer? Why or why not?

5. Would NOAA get lost in an expanded Interior department? Would its voice be muted? Why or why not?
6. How would moving to Interior affect NOAA's role? How might this affect NOAA's funding?

Step 2:

Take your opinions from questions 1 to 6 and write a one-page policy paper. This should explain which department NOAA should be a part of. Include an introductory paragraph to explain your policy. Follow the format outlined below.

- Paragraph 1: What is the question?
- Paragraph 2: What are the issues surrounding the question?
- Paragraph 3: What is your policy on these issues, and why?

QUIZ ★★

1 How many scientists and engineers work at NOAA?

2 Which U.S. president signed NOAA into being?

3 Which shipwreck was named the first NOAA national marine sanctuary in 1975?

4 Which 2019 hurricane caused more than $1.6 billion in damages in the United States?

5 Where is the Furnace Creek weather station?

6 How many people were killed in the Great Galveston Storm of 1900?

7 Which well-known figure was the great-grandfather of Alexander Dallas Bache?

8 Which Japanese nuclear power plant was severely damaged by a tsunami in 2011?

9 Which NOAA career involves flying aircraft into large storms?

10 What does a radiosonde measure?

ANSWERS

1. Almost 6,800 2. Richard Nixon 3. USS *Monitor* 4. Hurricane Dorian
5. Death Valley, California 6. Between 6,000 and 12,000 7. Benjamin Franklin
8. Fukushima Daiichi 9. Hurricane hunter 10. Temperature, humidity, and atmospheric pressure

KEY WORDS

aeronomy: the study of Earth's upper atmosphere

autonomous: self-operating

bycatch: unwanted fish and other marine creatures trapped by nets during fishing for a different species

constitution: a country's basic laws, which state the rights of the people and the powers of the government

ecosystems: organisms interacting with each other and their local environment

environment: the natural world as a whole, or the surroundings in which animals or plants live

executive branch: the U.S. president, vice president, and cabinet

executive order: a rule or order issued by the president, having the force of law

independent agency: a U.S. government agency outside of the executive branch

judicial branch: the judges and justices who interpret laws and decide if they are constitutional

legislative branch: the U.S. Congress, composed of the Senate and the House of Representatives

meteorologist: weather forecaster

regulations: rules set up by governments that determine how something is to be done

salinity: saltiness of water

seismology: the study of earthquakes

supercell: a large, slow-moving weather system that produces severe thunderstorms, hail, and tornadoes

sustainable: conserving an ecological balance by avoiding the depletion of natural resources

thermodynamics: the branch of physics that deals with relationships between heat and other forms of energy

tsunami: a large wave caused by an earthquake

INDEX

SUPPLEMENTARY RESOURCES

Click on the plus icon found in the bottom left corner of each spread to open additional teacher resources.

- Download and print the book's quizzes and activities
- Access curriculum correlations
- Explore additional web applications that enhance the Lightbox experience

LIGHTBOX DIGITAL TITLES
Packed full of integrated media

VIDEOS

INTERACTIVE MAPS

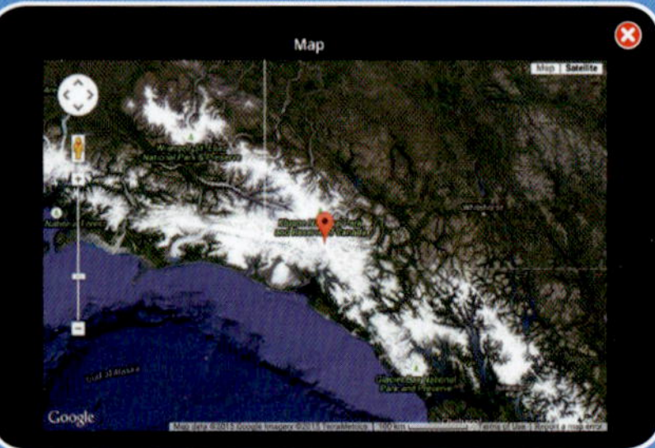

WEBLINKS

SLIDESHOWS

QUIZZES

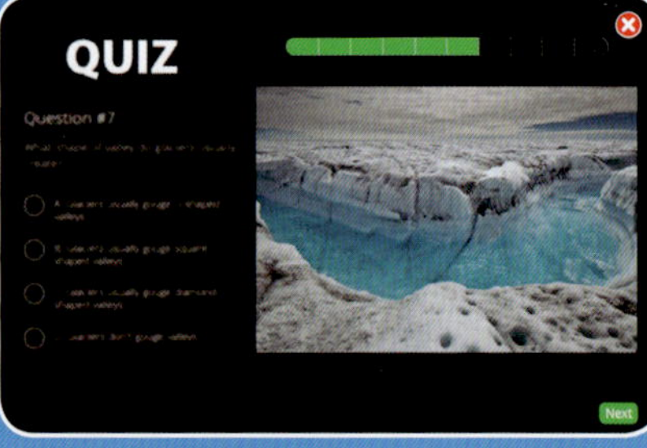

OPTIMIZED FOR

- ✓ TABLETS
- ✓ SMARTBOARDS
- ✓ COMPUTERS
- ✓ AND MUCH MORE!

Published by Lightbox Learning Inc.
276 5th Avenue, Suite 704 #917
New York, NY 10001
Website: www.openlightbox.com

Library of Congress Cataloging-in-Publication Data

Names: Hicton, Douglas, author.
Title: National Oceanic and Atmospheric Administration/ Douglas Hicton.
Description: New York, NY : Lightbox, [2022] | Series: Power, authority, and governance | Includes index. | Audience: Grades 4-6
Identifiers: LCCN 2021033740 (print) | LCCN 2021033741 (ebook) | ISBN 9781510558847 (library binding) | ISBN 9781510558854
Subjects: LCSH: United States. National Oceanic and Atmospheric Administration--Juvenile literature. | Oceanography--Research--United States--Juvenile literature. | Marine biology--Research--United States--Juvenile literature. | Hydrology--Research--United States--Juvenile literature. | Weather forecasting--United States--Juvenile literature.
Classification: LCC GC21.5 .H53 2022 (print) | LCC GC21.5 (ebook) | DDC 551.460973--dc23
LC record available at https://lccn.loc.gov/2021033740
LC ebook record available at https://lccn.loc.gov/2021033741

Printed in Guangzhou, China
1 2 3 4 5 6 7 8 9 0 27 26 25 24 23

012023
111121

Art Director: Terry Paulhus Project Coordinator: John Willis

Every reasonable effort has been made to trace ownership and to obtain permission to reprint copyright material. The publisher would be pleased to have any errors or omissions brought to its attention so that they may be corrected in subsequent printings.

The publisher acknowledges Alamy, Getty Images, and Shutterstock as the primary image suppliers for this title.